The Hopkins Method for Raising Queens in your Small Apiary

By Joe Conti

Northern Bee Books

The Hopkins Method for Raising Queens in your Small Apiary

Published 2021 by
Northern Bee Books,
Scout Bottom Farm,
Mytholmroyd,
West Yorkshire
HX7 5JS (UK)
Tel: 01422 882751 Fax: 01422 886157
www.northernbeebooks.co.uk

ISBN 978-1-914934-25-4

Design and artwork DM Design and Print

The Hopkins Method for Raising Queens in your Small Apiary

By Joe Conti

Figure 1. Hopkins setup.

Introduction

Beekeeping presents its challenges, and one of the most troublesome is finding colonies with no queens. There are queen suppliers who are more than willing to sell you some queens, but the prices of these "royal highnesses" seem to be increasing each year. These purchased queens also come from large-scale commercial queen rearing operations, and the source of their genetics is largely unknown given the large mating yards with hundreds of thousands of drones that are mating with those queens. It could be a risky proposition to spend a lot of money on queens of questionable origin.

Beekeepers can raise their own queens, and they have a wide range of choices for this. Most of the techniques are usually designed for raising many queens, and involve intricate tools and procedures. Some of the equipment beekeepers purchase to raise queens include grafting tools, cell bars, plastic cell cups, plastic cell protectors, candy caps, and queen cages. There are even complete queen rearing kits that include everything you need plus instructions. Some of the common methods associated with queen rearing are the Jenter, Nicot, and Doolittle systems. The bee supply catalogs offer all the equipment needed for these methods at a nominal cost, and they all are successfully proven. Some of the methods involve grafting young larvae into queen cups and require a considerable skill set, while others are non-grafting, only requiring the special equipment for queens to lay in.

The other option is quite simple: let hives requeen themselves naturally. But supposing you have a hive or two that is superior to the rest of your stock, and you want to use eggs from those queens to requeen your inferior stock so all of your hives have a chance at becoming superior. A beekeeper can remove frames of eggs/young larvae from the superior colonies and place them in each of the inferior colonies…but knows it comes at a cost to the superior colonies you are stealing frames from. Is there another way to use eggs/young larvae from your superior colonies without negatively affecting their production? As it turns out, yes…..and there is an often overlooked and simple method to producing your own high quality queens from your own stock: the Hopkins Method. Best of all, this method requires no purchasing of queen rearing materials or kits, nor learning curve and skills required of grafting.

THE ILLUSTRATED AUSTRALASIAN BEE MANUAL AND COMPLETE GUIDE TO MODERN BEE CULTURE IN THE SOUTHERN HEMISPHERE

HOPKINS, ISAAC

Figure 2. Australasian Bee Manual, available on Amazon.com

What Is The Hopkins Method?

The Hopkins Method is the complete removal of a frame of eggs/young larvae from a selected colony and placing it horizontally in a queenless colony.

History

The Hopkins Method was first described by Isaac Hopkins in 1911 in a publication called The Illustrated Australasian Bee Manual And Complete Guide To Modern Bee Culture In The Southern Hemisphere (Figure 2). It is still available today. Isaac Hopkins was the Chief Apiarist to the New Zealand Government who made keen observations and performed various experiments in his quest to modernize beekeeping practices in New Zealand. In the Manual, Hopkins gives credit to an Austrian beekeeper who first tried and described it (no name given), but credits himself as being the first person to perform a trial on the technique in his part of the world. The technique is sometimes referred to as the Case Method (mostly in the United States) after being presented by H.L. Case, but it seems only fitting that Isaac Hopkins get full credit as he was the first person to experiment with it and report on successful results.

If the Hopkins Method was utilized by beekeepers worldwide, it certainly wasn't popularly discussed in beekeeping publications. In fact, decades passed until an article by Jerry Hayes in the American Bee Journal in 1984 and again by him in 1991 brought it into focus. Thirty more years passed before an article I wrote in the 2021 American Bee Journal fully described the method as I used it in my apiary, complete with photos (Figure 3). With this latter article, it was hoped that beekeepers would become more familiar and comfortable with using this simple method of raising queens.

NEED A QUEEN? TRY THE HOPKINS METHOD

by JOE CONTI

Beekeepers have a wide range of choices for raising queens. There are books written on the subject and oodles of items one can buy. Just do a Google search on "queen rearing" and you will discover all sorts of interesting techniques and products. Many, however, are designed for large-scale queen rearing, and involve intricate tools and procedures. Some of the equipment beekeepers purchase to raise queens include grafting tools, cell bars, plastic cell cups, plastic cell protectors, candy caps, and queen cages. There are even complete queen rearing kits that include everything you need plus instructions. So, what if you're

a backyard hobbyist and you only need enough queens to supply your small-scale apiary? Is it necessary to purchase all these materials? Should you just purchase queens of questionable origin and condition? Or is there a better way? It turns out that there is an often overlooked and simpler way to produce high quality queens in your own apiary, using your own choice of stock: the Hopkins Method.

The Hopkins Method originated with a Mr. Isaac Hopkins in New Zealand in 1911 (it later was called the "Case Method" in the U.S.). The Hopkins Method was described by Jerry Hayes in the American Bee Journal in 1984 and again in 1991. Maybe after 30 years in the literature, it is time for beekeepers to revisit this simple method of raising queens. And I'm convinced that if I can do it, any beekeeper would be capable of the same.

Now, prepare yourself for how easy this is. The basic premise is to choose a hive that you feel has the best characteristics that you would want to raise queens from, thereby keeping those genetics in your apiary. The steps below are a simplified version of the method, so I would suggest checking online for subtle nuances that could add to your success.

1 Pick a strong colony that you know has enough bees, along with a great laying queen, to increase the likelihood of having lots of nurse bees. It is the nurse bees that will be important in the construction of queen cells and the feeding of royal jelly to the developing queens. I will refer to this as the RECIPIENT colony.

2 Remove the queen from the RECIPIENT colony (use her for whatever purpose you need).

3 Pick another strong colony that you feel has the right characteristics that you would like to maintain, such as gentleness, hygienic behavior, a good laying pattern, low mite counts, etc. I will refer to this as your DONOR colony.

4 From this DONOR colony, remove one frame of eggs/newly hatched larvae from the center of the colony. (NOTE: You could also just use a frame from the RECIPIENT colony if it is your best choice.)

5 Place the frame of eggs/newly hatched larvae horizontally across the top of the uppermost

Ten frame deep hive body (RECIPIENT hive) with 2½" shim above a few frames just for illustration. Notice the 1¼" chiseled notch and the plastic piece attached to the far end board of the shim. A notch and plastic piece are on the opposite end as well.

Horizontal frame inserted into shim. This frame would represent the DONOR hive that has characteristics you want to raise queens from.

Figure 3. Published article on Hopkins Method, American Bee Journal, May 2021, pp. 551-553.

8

So, let's get into the 'nuts and bolts' of how this method works. There are a few prerequisites before getting started: you must have an excellent breeder queen in a strong and successful colony (now referred to as the DONOR COLONY); a second strong colony with copious amounts of nurse bees that will build the queen cells (now referred to as the RECIPIENT COLONY); and remaining colonies that will need requeening, or nucleus (nuc) boxes should you decide to create new colonies. It should also be done in late spring/early summer through the late summer/early fall only, when colonies are bountiful and if there are plenty of drones present. It would also be helpful to augment the drone population by adding a drone frame (green in color, available commercially) to each superior hive (explained more in detail later).

The one piece of equipment you would need to assemble yourself, and is easily accomplished, is a shim box that will hold a frame of bees in a horizontal position (see Figures 4, 5, 6, 7, and 8 for illustration purposes). Once this is made, you can store it to be reused every year.

Figure 4. Shim for placing DONOR frame horizontally. Measurement in this particular shim was 16¼ inches wide, by 20 inches long, by 2½ inches tall (but this latter value could vary to your liking). It has notches chiseled out on each side to fit the ends of the top bars of the horizontal frame. You will need something at the other (bottom) end of the frame to level it out. I used a plastic wire shelf holder screwed in on the left shim, but you could use a strip of wood for same leveling result as seen on the right shim. The depth of the chiseled notch for the top bar to slide into was 1¼ inches, but this depth will vary depending on how tall a shim you decide to use. the main point, however, is to make a notch deep enough that leaves you with a space approximately 1½ to 1¾ inches between the horizontal frame (bottom surface with eggs/young larvae) and the surface of the top bars of the 10 frames below. This space is necessary for the development of large queen cells similar to the swarm cells bees make at the bottoms of frames during swarming season.

Figure 5. Horizontal DONOR frame sitting in shim.

Figure 6. Shim sitting on deep brood box (frames added for illustration purposes)

Figure 7. DONOR horizontal frame being placed in shim

Figure 8. When the shim with a horizontal frame in place is situated atop a regular hive body box, there should be at least a 1½ inch space between bottom of DONOR horizontal frame and top bars of the frames in the box below.

Scenario

Now let's assume you have between 10-20 hives, and you are a beekeeper that likes to requeen colonies with vigorous, young, healthy queens each year. The choices are to buy queens at $30 apiece, or raise your own. And let's assume you decided to raise your own queens because buying them has some drawbacks (discussed later). Your choices of raising queens come down to 2 options: (1) buying the queen-rearing materials, or (2) letting your colonies requeen themselves. Upon examining the quality of your hives, you notice that there are very few, and possibly only one colony, that you would consider "high-quality", in other words, possessing a queen that is a prolific layer, bees that are gentle, relatively disease-free, etc. And you are aware that just removing a queen from each hive, letting all the colonies requeen themselves, is essentially perpetuating the "inferior" genes in the lower quality colonies. It occurs to you that you could remove several frames of eggs/young larvae from your one, high-quality colony and place them in the queenless lower quality colonies, and this would accomplish your goal…but at the expense of the high-quality colony. You decide that a better way to accomplish this feat is to use the Hopkins Method, whereby you remove only 1 frame from your high-quality colony, and raise multiple queens on that one frame in another colony. Your beekeeper friend tells you, "Hey, why not just remove your high-quality queen and let that colony raise many queens in the same hive?" Seems like a logical suggestion, but your better sense tells you that you'd rather not disrupt an already highly functioning colony, in addition to wondering if the colony would choose older larvae to rear queen cells with, thus increasing your chances of collecting inferior queens. By virtue of having only the youngest larvae available in the Hopkins Method, you are insuring a very high probability of creating superior queens.

Procedure

Step 1. Choose Your Superior Colony (DONOR COLONY)

Find a strong colony that has a history of exhibiting the best characteristics you can find within a hive as it relates to gentleness, hygienic behavior, laying pattern, prolific layer, disease resistance, etc. (Figure 9).

Figure 9. DONOR COLONY, chosen because a gentle colony and queen prolific layer. Frame of capped brood removed for illustration. This brood frame will be taken out permanently and replaced with an empty drawn comb as shown in Figure 11.

Step 2. Choose Your Queen-Rearing Colony (RECIPIENT COLONY) (Figure 10).

Figure 10. RECIPIENT COLONY, chosen because populous colony with much brood, hence many nurse bees.

Now pick a strong colony that has plenty of bees, and plenty of eggs and brood, as this colony will have a sufficient amount of nurse bees that will be able to construct queen cells and feed royal jelly to the developing queen larvae. Some beekeepers refer to this colony as the cell-builder hive.

Once Steps 1. and 2. are determined, you can now move on to the actual manipulation steps as described below.

Step 3. Remove A Frame From The Middle Of The DONOR COLONY (see Figure 9) And Place An Empty Frame Of Drawn Comb In Its Place (Figure 11).

If there are any other frames with empty cells in the DONOR COLONY, it may be best to remove these as well. The idea is to have just one frame that the DONOR COLONY queen MUST lay eggs in because she has no other choices. Also, make sure that the queen is in the brood nest area and can't wander to any upper boxes (use an excluder if necessary). Leave the empty frame in the hive for about 3 days, as this will give the queen enough time to fill out at least one side of the fame. The queen may have filled the frame out before 3 days, which is fine, but do not wait any longer than this as the eggs will start hatching on Day 4…the day you would like to have this frame in place in the queen-rearing RECIPIENT COLONY.

Also, when choosing a drawn frame with empty cells to place in the middle of the DONOR COLONY, try to stay away from frames of pure wax unless you have a way of supporting the wax in the center parts of the frame (which I'm sure can be done with reinforced wire); just realize that heat from the hive and outside temperatures will cause the foundation to sag when laid horizontally in the queen-rearing RECIPIENT COLONY, especially with the weight of the bees.

Figure 11. Empty drawn comb frame being placed into DONOR COLONY in order for superior queen to lay eggs in.

Step 4. Remove The Queen And Frames With Eggs/young larvae From The RECIPIENT COLONY Rendering It Queenless

This step is done approximately 1 or 2 days before your decision to place the DONOR COLONY frame with eggs/young larvae in the RECIPIENT COLONY. This will give the RECIPIENT COLONY enough time to realize they are queenless, and they will commence to starting new queen cells on the DONOR COLONY frame once it is inserted. During the removal of the queen, it would be wise to remove any frames with eggs/young larvae with her as well, since you don't want the RECIPIENT COLONY to start raising queen cells from their own queen within their hive. It is OK to leave older larvae as these would typically not be chosen for creating queens. Of course, capped brood can be left as well since they will become an additional source of nurse bees once the baby bees emerge. This is a good time to create a new colony with a small split, or nuc, by combining the removed queen and the older larvae (and some capped brood) with an added frame or two of food (honey/pollen). It is important to note that if you are removing frames from the RECIPIENT COLONY, try to brush off the bees that are adhering to the brood frames, as these are nurse bees which are abundantly needed in the RECIPIENT COLONY for them to properly raise the newly desired queen cells. If you are needing bees to add to your new split, it is preferable to take bees from neighboring colonies, being careful not to brush off a foreign queen into your newly established queenright hive.

Step 5. Place The DONOR COLONY Frame Of Eggs (or day-old larvae if some hatched), Now Horizontally In The Shim, Across The Top Of The Brood Box Of The RECIPIENT COLONY.

Before this is done, Hopkins noted (as other beekeepers) that if eggs are not spaced out some, that queen cells many times become bunched together, making separating them somewhat difficult when it comes to cutting them out. While logical given this bee behavior, I don't believe this is absolutely necessary, as you'll always find individual queen cells to cut out, and you may have more queen cells than you need anyway. However, for the purpose of describing the Hopkins Method, he recommended that every 4th row should be left with eggs, with all 3 rows in-between scooped out; and 2 out of every 3 eggs in the standing rows destroyed (Figure 12). This may be the most tedious process of the Hopkins Method, and other beekeepers after him have come up with alternative "tic-tac-toe" patterns. The point is, separate eggs as you wish so as not to encourage the bees to build queen cells attached to each other. Now for all practical purposes here, if you are requeening a colony, you may find it acceptable to put an attached couple of queen cells (removed in tandem) in your hive; in fact, it is wise indeed to have 2 queen cells in a hive anyway, to increase your chances of a successful queen emergence compared to having only one cell. In addition, as mentioned previously, if you only need a small amount of queen cells, this method will produce a lot of queen cells, some attached and others unattached…just collect the unattached ones; or, if you're adventurous and like a challenge, try separating attached queen cells. The tedious task of destroying rows and columns of eggs is not necessarily critical to success using the Hopkins method.

Figure 12. Rows and columns of eggs/young larvae destroyed to minimize queen cells becoming attached to one another. The illustration shows a more generous leaving of viable cells as compared to the Hopkins method of destroying nearly all rows and columns until only one egg/young larva is isolated by itself. I found his technique difficult to achieve without creating a mess, and probably unnecessary if you are trying to create just a few queen cells. However, his technique may be more easily accomplished if the wax on the frame was newly formed and thus more pliable. There are visibly older larvae in the cells closer to the bottom of the frame in this figure, which should be removed entirely so the bees don't choose them to rear queens (usually unlikely, but a safeguard against that happening).

The process of placing the DONOR COLONY frame HORIZONTALLY into the RECIPIENT COLONY is illustrated (Figures 13, 14). Now Hopkins himself did not use a shim box prepared as described herein, but simply laid a totally empty frame across the top bars of the brood frames, proceeded by laying the DONOR COLONY frame of eggs/young larvae on top of that. This would have necessitated his use of a larger shim than described in this booklet, as the frame would not fit within a standard hive (unless the top bar ends were cut off, rendering the frame useless after that point). He also may have short-changed the amount of space the nurse bees could have had to create large queen cells. It is recommended here to make sure there is at least 1 ½ inches of space for the creation of queen cells no matter what technique you use to lay the DONOR COLONY frame horizontally.

Figure 13. DONOR COLONY frame in shim placed on top of brood (deep) box of RECIPIENT COLONY.

Figure 14. RECIPIENT COLONY with shim above the brood nest, and situated between Medium and Deep hive bodies.

Step 6. Wait 10 Days Until Capped Queen Cells Are Observed (Figures 15, 16)

Figure 15. Capped queen cells on DONOR COLONY frame
(note: this was the side that was facing downward over the top bars of the
RECIPIENT COLONY, now flipped over for viewing).

Figure 16. Better view of capped queen cells.

Remember, you put 3-day old eggs in the RECIPIENT COLONY, and 10 days later is 13 days total. A queen will emerge in 16 days after an egg is laid, so you have 3 days to spare.

Step 7. Remove Capped Queen Cells At 10 Days With A Sharp Tool

Find the sharp tool of your choice (I use a sculpting blade; Figure 17) and get to the very base of the queen cells, slicing carefully until it is free (Figure 18). While this works best with all-wax frames, it will work with plastic frames, but you have to be extra careful while cutting as you don't want to damage the developing queen. Cutting right up against the plastic frame will work, but in many cases, there will be a hole at the base, revealing the queen late larva/pupa inside (Figure 19)... but not to worry. Either gently squeeze the wax at the end to form a closed seal, or simply take a thin piece of beeswax and plug up the hole; it will work even better if the wax is slightly warmed and malleable (Figure 20). This will create enough of a seal for protection. And even if it didn't, placing a queen cell with a hole is not tantamount to failure... the bees will plug it up themselves.

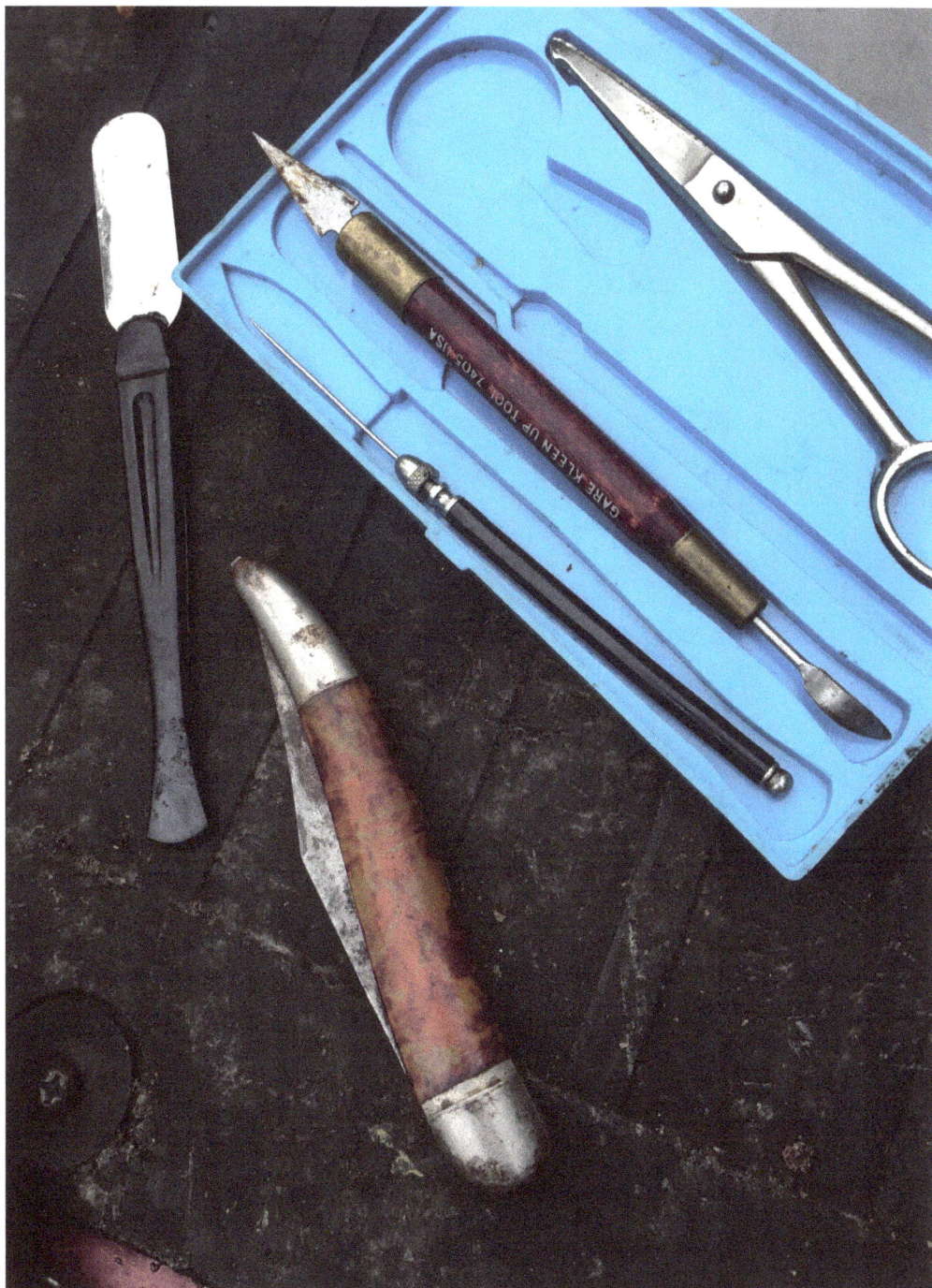

Figure 17. Some instruments you may want to have handy when cutting out queen cells. The sculpting tool is centered in the figure. It might be good to have some extra wax available for queen cells that are acidentally opened while cutting.

Figure 18. Cutting queen cell out with sculpting tool.

Figure 19. Base of queen cell accidentally exposed while cutting.

Figure 20. Base of queen cell plugged with soft wax.

Step 8. Place Queen Cells In Your Queenless Colonies Or Nucs (Figures 21, 22).

Figure 21. Wedge queen cell between frames of new colony you wish to requeen.

Figure 22. Bees getting to work fixing the base of the queen cell to their liking, and securing it better to the sides of the frames.

And you're done! Again, your new colonies should have been rendered queenless a few days prior to inserting the new queen cells. The new queen cells can be wedged in-between the top bars, or placed anywhere within the center of the brood nest with whatever adhering mechanism you desire (gently pressing into a space on the comb; toothpicks; etc.).

The Problem of Unknown Drones and the breeding of your Superior Virgin Queens

Finally, if you are raising 20 or more queens, make sure they have plenty of drones to mate with once all those virgin queens take flight. If a dozen or more drones are necessary to mate with one queen, you will need hundreds of drones available at any one time. A good idea would be to place some drone frames (green-colored, available commercially) within the superior colonies of your apiary, and make an attempt to destroy drone comb in the less than desirable colonies. The goal here is to saturate the drone congregation area (DCA) with as many drones as possible from your selected best colonies. Unfortunately, beekeepers have no control over where their virgin queens go to mate. It is known that queens tend to fly past their own DCA to mate with drones in other DCAs, thus avoiding an inbreeding situation. This is not done consciously, but it is more of an innate behavior and possibly related to energy reserves on the part of the bees (queens fly far, mate quickly, fly back, conserving energy; drones hover for long periods in their congregation areas, but close to the colony to conserve energy). This might suggest that the Hopkins Method is a waste of time, since the question arises as to how can one achieve superior colonies when the superior queens are out mating with unknown drones. Understand that this question could also be asked of any large-scale queen breeding operation. While this is certainly true in all matters related to raising queens, the Hopkins Method, coupled with drone selection, may have a distinct advantage over commercially raised queens. For one, at least in the Hopkins Method you are choosing eggs of a superior queen; and if you combine this with culling out undesirable drones in inferior colonies while promoting desirable drones in superior colonies, as mentioned earlier, you would be strengthening the entire gene pool of your area. This is a good thing for, if queens from neighboring colonies fly to your apiary DCA (which is close to your hives), they would be mating with higher quality drones because of your superior queens and resulting drone selection process. This in turn improves the quality of the neighboring colonies, including the drones. When your virgin superior queens fly to these far off DCAs, they will at least have higher quality drone populations to mate with, thanks to your efforts at the home apiary. This then becomes an excellent recipe for creating, as well as sustaining, the best colonies, not only in your own apiary, but in other nearby apiaries as well. Keep in mind that the practice just described is not necessarily done in a typical commercial queen breeding operation, where queen and drone numbers, not necessarily qualities, are critical. When commercially raised queens mate with drones from a commercially prepared drone mating yard, the quality of these drones was not necessarily controlled as compared to what you can do within your own smaller apiary. Another factor to consider is, what if there are no other DCAs in the nearby vicinity and queens end up mating with drones from their own apiary? Isn't inbreeding a bad thing, as inbreeding could lead to undesirable colony characteristics over time? If an apiary is large enough, inbreeding may not be as big a factor, and in some cases may actually be a good thing initially as certain high-quality traits may become fixed in the population. To avoid the increasing chances of undesirable inbreeding characteristics, it might be wise to obtain (and introduce) a known superior queen from outside your apiary every few years to diversify the genetics as an anti-inbreeding measure.

In retrospect, the Hopkins Method not only will allow you to produce great queens, but also great colonies if you combine the method with drone selection. It is a simple solution to raising superior quality queens in your own backyard without having to invest in a whole lot of materials, or gain any sort of expertise. The steps outlined previously appear lengthy in explanation, but they are actually easy to follow because much of it is intuitive; and once you get the 'know-how', the technique will become easier to accomplish with each succeeding attempt. You would think that a technique as simple as this would end up as a common tool in the beekeeper's handbook, yet it has been ignored. Perhaps with some new publications such as this, the Hopkins Method will be revived after a 110 year hiatus, and be discussed commonly in beekeeping forums everywhere.

THE ILLUSTRATED
Australasian Bee Manual

AND COMPLETE

GUIDE TO MODERN BEE CULTURE

IN THE

SOUTHERN HEMISPHERE.

———

By ISAAC HOPKINS, Auckland, New Zealand.

(Late Chief Apiarist to the New Zealand Government.)

———

WITH WHICH IS INCORPORATED THE

"New Zealand Bee Manual"

REVISED AND MOSTLY RE-WRITTEN
BY THE AUTHOR.

———

FIFTH EDITION.

82 ILLUSTRATIONS.

———

WELLINGTON, N.Z.

1911

GORDON & GOTCH

Published article on Hopkins Method, American Bee Journal, May 2021, pp. 551-553.

Editorial, Notices, &c.

PROMINENT BEE-KEEPERS.

MR. ISAAC HOPKINS.

We have pleasure in presenting this week the portrait of Mr. Isaac Hopkins, late Government expert, and a pioneer bee-keeper of New Zealand.

Although we believe Mr. Hopkins was born in England, for upwards of Manual," which awakened an interest in bee-keeping among the settlers of the colony, so that a second edition was called for in 1882. When this edition was exhausted Mr. Hopkins had to consider the greatly altered circumstances, and in issuing the third in 1886 he made it suitable to the new conditions and the advances made up to that time. This was published as "The Illustrated Australasian Bee-Manual," a book of some 300 pages, with which the previous work was incorporated, and in the compilation of which he was assisted by Mr. T. J. Mulvany, of Bay

MR. ISAAC HOPKINS.

thirty years he has been an energetic and useful bee-keeper in New Zealand, and the progress of the industry in that colony is mainly due to his initiative, for he was the first to introduce modern bee-culture into Australasia. Previous to 1878 he had been trying different forms of frame, but in that year he became acquainted with the "Langstroth" hive, and has since then advocated it as the one best suited to New Zealand. As bee-keeping was being taken up the want of a manual suited to the requirements was felt, so in 1881 he published an excellent book of instruction, "The New Zealand Bee-

View Apiary, Katikati, the fourth edition being reached in 1904. Mr. Hopkins was manager for Mr. J. C. Firth at the Matamata Apiary, Waikato, Auckland, for some time, and an incident at this apiary, in 1883, when there were about 200 hives, some of them two and three stories high, and in one case five stories, is worth recalling. This last was a swarm of the current year, and stored 240 lb. of honey, besides building out fifty sheets of comb-foundation. The parent colony gave another swarm, which produced 210 lb. This, together with about 100 lb. it had itself, made for the colony and its produce

460 lb. surplus honey—a result truly wonderful.

In 1891 we find him trading as Hopkins and Co., honey merchants, for which firm he compiled the pamphlet, "Honey, the Natural Sweet for Human Food." In 1883 he had supplied an article on bee-keeping to "Brett's Colonial Guide," which was enlarged in the second edition in 1897. He edited the *New Zealand and Australian Bee Journal* during 1883 and 1884, the two years of its circulation, and this brought him into communication with bee-keepers in all parts of the colonies.

In 1888, when secretary of the New Zealand B.K.A., he assisted in drafting a Bee-Pest Bill for the New Zealand Parliament, and though compelled by ill-health to abandon the scheme for a time, he never ceased to cherish it, and by dint of hard work up and down the country, during which he formed various bee-keepers' associations, he had the satisfaction to see the New Zealand Apiaries Act come into force on September 14, 1907. This Act is the most progressive and at the same time the most practical yet in force in any country. It grants no compensation for the compulsory destruction of diseased hives and stocks, and renders the keeping of the honey-bee in any dwelling except hives with movable frames an illegal and highly-penalised act. There could be no greater tribute to the energy and persuasive power of Mr. Hopkins, as also to the good sense of New Zealand bee-keepers, than the fact that the measure is popular from one end of the country to the other.

For the last few years Mr. Hopkins has been the New Zealand Government expert, with assistant experts under him. He set up the Exhibition Apiary at the Great International Exhibition in Christchurch in 1906-7, in which he was assisted by Miss Livesay, who holds the B.B.K.A. expert certificate. He has also established two Government bee-farms, and has produced for the Department of Agriculture two bulletins; that entitled "Bee-Culture," published in 1907, has this year reached the third edition.

Travelling in so large a colony, lecturing and forming bee-keepers' associations, must be hard work, and Mr. Hopkins has just retired from the Government service to his well-earned rest at the age of seventy-two. Neither his record nor his portrait that we are pleased to reproduce encourages the idea that he will care to be an idle man, and we trust for years to come he may flourish as a firm supporter of the craft.

The prospects for commercial bee-keeping in New Zealand are now very bright. Many of the old bee-keepers had nearly given up bee-keeping because there was no control over the careless and wilful box-hive men, who were propagating disease and causing continuous trouble and loss. With the advent of the present policy and the Apiaries Act bee-keepers took heart again, and the industry has since gone ahead by leaps and bounds, being now established on a thoroughly sound and commercial basis.

REVIEW.

Nature Through the Microscope. By William Spiers, M.A., F.R.M.S. (published by Robert Culley, 25, City Road, London. Price 7s. 6d. net).—In this book the author has aimed at supplying such information to lovers of Nature as will make their excursions in the country and by the seashore interesting and instructive. It is written in as simple language as the subject-matter permits of, and with the help of drawings and photographs it is hoped that the non-microscopical reader will be able to participate in the pleasures which are enjoyed by the microscopist. Only those who are in the habit of using the microscope can have any idea of the marvels hidden beyond the reach of ordinary vision, and of the wonderful beauty and some of the most perfect forms of the minutest structures that one comes across. How different from the work of man, which under the microscope shows its coarse imperfections, is that of Nature, which reveals its wonderful perfection the more it is magnified. The Rev. W. Spiers is known as a writer on microscopical objects, and in the work before us he has produced a book which should become popular, for he has avoided technicalities as far as possible while adhering to the methods of text-books. It begins with a chapter on the beautiful in little things, being descriptions and illustrations of desmids, which lie at the very first rung of the ladder of botanical classification. The following chapters advance step by step from the bottom to the top of the biological ladder, making it a pleasant introduction to botany and natural history in general. The 355 pages are divided into thirty-eight chapters, and bee-keepers will naturally be most interested in those parts relating to bee-structures, which are suitably illustrated. There are ten coloured plates, and in all ninety-nine, containing 300 drawings and micro-photographs of just such things as one wants to know something about. We have much pleasure in recommending the book to our readers as one suitable for a Christmas present for anyone wishing to know something about the microscope, the wonders it reveals, and the preparation of suitable objects for examination.

www.ingramcontent.com/pod-product-compliance
Lightning Source LLC
Chambersburg PA
CBHW040155200326
41521CB00020B/2610